SIMPLE APPLIANCE REPAIR

Other Books by Ross R. Olney

SIMPLE BICYCLE REPAIR AND MAINTENANCE
SIMPLE GASOLINE ENGINE REPAIR

SIMPLE APPLIANCE REPAIR

ROSS R. OLNEY

ILLUSTRATED BY MARY ANN DUGANNE

DOUBLEDAY & COMPANY, INC., GARDEN CITY, NEW YORK   1976

Library of Congress Cataloging in Publication Data

Olney, Ross Robert, 1929–
  Simple appliance repair.

  1. Household appliances, Electric—Maintenance
and repair. 2. Gas appliances—Maintenance and
repair. I. Title.
TK7019.046   683′.83′028
ISBN 0-385-11098-7
Library of Congress Catalog Card Number 76–2810

COPYRIGHT © 1976 BY ROSS R. OLNEY
ALL RIGHTS RESERVED
PRINTED IN THE UNITED STATES OF AMERICA
FIRST EDITION

CONTENTS

FOREWORD vii

1
GENERAL REPAIR PROCEDURES 1

2
DISHWASHERS 13

3
WATER HEATERS 25

4
WASHING MACHINES 37

5
CLOTHES DRYERS 49

6
REFRIGERATORS AND FREEZERS 57

7
TOASTERS 71

8
COFFEE MAKERS 77

9
OTHER SMALL APPLIANCES 83

    ELECTRIC IRONS   83
    BROILERS AND ROTISSERIES   85
    VAPORIZERS AND BOTTLE WARMERS   86
    ELECTRIC FRYING PANS AND COOKERS   88
    CROCK POTS   90
    SPACE HEATERS   91
    HAIR DRYERS   94
    GARBAGE DISPOSERS   94
    VACUUM CLEANERS   96
    ELECTRIC CAN OPENERS   97
    BLENDERS AND MIXERS   98

INDEX   101

FOREWORD

There is a story that still circulates today about a British radar instrument in World War II and the repair manual that was issued with it. Repairmen were given detailed lists of instructions, step-by-step procedures to follow, and guides on what to do if anything went wrong with the unit.

The final order in the manual, to be carried out if all else failed, was "KICK HARD!"

Let's get one thing clear from the beginning. There isn't a thing wrong with a jiggle, a bump, a wiggling of the door, or a little adjustment of a dial if a home appliance suddenly doesn't work. Maybe a hard kick should be reserved until all else fails, but there is no need to panic if an appliance stalls.

Nor is there any need to rush for the telephone to call an expensive repairman. The fact is, more than twenty-five per cent of all service calls on home appliances (that's one out of every four calls) need *not* have been made. Not because the appliance owner could finally have figured out a very complicated repair procedure himself, but because the repair procedure was so simple that *anybody* could have done it on the spot. A plug out of a socket, a switch turned the wrong way, a fuse blown, or some other very uncomplicated matter could be the problem.

Wiggling a switch might help. Bumping a unit that might be clogged could help. A little adjustment of a dial could get things going again. Moderation in all such matters is important, but just because an appliance has stopped doesn't mean that you are in deep trouble. Not yet.

This book is designed to be used *before* you place any calls for help and this Foreword and Chapter 1 are designed to be read before you refer to specific problems in the book.

If an appliance fails, the following list of suggestions might get it working again. Some of the points are basic, but any one of them could solve the problem.

1. Be patient. Do not hurry, even if you have a refrigerator full of food and company knocking at the door. Haste can make a simple problem more complicated, so take your time.

2. Be sure the electric cord is plugged in. Don't just look at it—take it out and replug it. Sometimes a slight spreading of the prongs will make a better contact.

3. The electric outlet could be faulty. Unplug the appliance and plug in a small electric lamp that you know is working. If it lights, you have electricity at the plug.

4. If electricity is not reaching the plug serving the appliance, check the fuse or circuit breaker. When in doubt, change the fuse; and if you are not certain of a circuit breaker, switch it off, then press it back toward the "on" position.

5. If you have no power to the plug serving the appliance and no evidence of blown fuses or tripped breakers, be certain that electricity is available to the house. Check other circuits. Ask a neighbor. Sometimes power is off throughout the neighborhood.

6. Read the manual and papers that came with the appliance. Be certain the controls are set properly and that all buttons and switches are set properly.

7. Go over the checklists that often come with an appliance. Sometimes you can recognize the problem and make the correction.

8. Have you maintained the appliance according to the manufacturer's instructions? Sometimes a good cleaning will correct a problem.

If none of these points help and you have decided to get into the matter as described in one of the following chapters, keep this checklist in mind.

1. It is better to do nothing if your appliance is still under some warranty from the store or the manufacturer. You might try a call to the dealer where you bought it, describing the problem and asking if there is anything you can do. The dealer will want to know if you have looked for the obvious; then he might be able to guide you on the telephone.

Most companies, however, are very fussy about their warranties. Something as simple as turning a screw or removing an inspection plate can invalidate a guarantee; so be careful. If you have a service policy you have bought and paid for, you can go a little further into the problem, but if you have such a policy, why bother? Just call the man and let him worry. (The trouble with service policies is that at first they are generally not needed, and by the time you need them you have sometimes put in enough money paying for the policy to have bought a new appliance.)

2. If you are going to take things apart, keep close track of all parts. Some home repairmen align the parts, including specific nuts and bolts and screws, in the exact order they are removed. Many of the procedures described in this book do not require disassembly, but if something is removed, be sure you have a way of telling how and in what order it is to be reinstalled.

3. The procedures and checklists in the chapters that follow are generally arranged in logical order. For example, if a dishwasher won't fill, the first thing to check (after you make sure there is electricity and water available) is the filter screen, the next is the inlet valve, and so forth.

4. Try to understand exactly how a machine works before you try to repair it. Understand as much as you can about the insides of an appliance you are about to dismantle. The parts guides and exploded views generally provided by the manufacturer are invaluable aids to the home repairman, not only in understanding the inner workings but also in locating and ordering a specific part needed for repair.

5. There is more than enough electricity in a household wall plug to do serious damage, or worse, to your body. That same electricity is available at many places inside most appliances. Be *certain* to unplug the appliance from the wall before beginning any repair procedure. If the appliance must be on for testing or diagnosing, be certain you are dry and away from any water on the floor, and that the appliance and any tools you are using are well grounded.

Many professional repairmen go around looking as if their hands were cold; they work with one hand in their pocket, especially if they are diagnosing an electrical problem in a plugged-in appliance. There is a reason for this.

An electric shock, depending on many things, can be anything from an unpleasant tingle to fatal. One of the things that matters is how the electricity passes through your body. If, for example, it only passes from your hand to your shoulder and out, it is better than passing in one hand, through your chest, and out the other hand. So repairmen often do certain things with one hand safely out of the way in their pocket.

With good grounding (electricity *prefers* to get to ground the easiest, quickest way) and normal safety precautions, you should not have to worry about electric shock, but it doesn't hurt to be careful.

6. Be wary of sharp edges and sheet metal screws and other fasteners protruding on the inside. The outside of most home appliances is smooth and shiny and safe, but you are going to be inside for some procedures and the manufacturer quite logically didn't worry about that part of the machine in considering sharp edges and points.

7. Some machines have exposed pulleys and gears inside. Even if the appliance is off and unplugged, keep your fingers away from these parts. Moving something over here can grab up a finger over there in some larger appliances, and even some smaller ones have strong springs and levers that can pinch.

8. Be careful of loose clothing around working appliances. Keep your sleeves rolled up tight and your hair out of the way.

9. To be prepared for real emergencies (smoke boiling from a clothes dryer, water gushing from a washer, etc.), be sure you know exactly how and where to shut off the water to your house or apartment, the exact location and identification of each fuse or circuit breaker, and any gas shutoffs outside, probably at the meter. Once such a major emergency has been handled, you will have time to go back to item 1 on this list and be calm and patient.

10. Have something available to protect the finishes of both the appliance and your working surface. Sometimes it is easier to work on a washer, for example, if it is on its side on the floor. Place it on a blanket to keep from scratching everything. If you are working on a small appliance on top of a table, cover the table with newspaper.

SIMPLE APPLIANCE REPAIR

1

GENERAL REPAIR PROCEDURES

HELPFUL TOOLS

Refer to the illustration for a list of tools that should answer any problem you will encounter in simple appliance repair work.

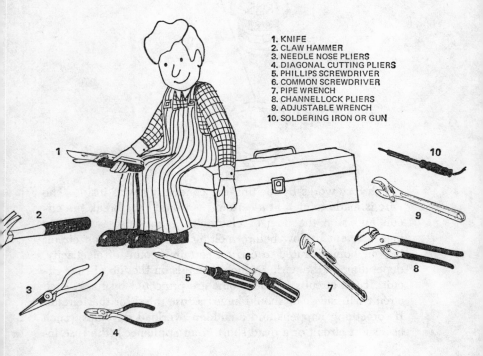

1. KNIFE
2. CLAW HAMMER
3. NEEDLE NOSE PLIERS
4. DIAGONAL CUTTING PLIERS
5. PHILLIPS SCREWDRIVER
6. COMMON SCREWDRIVER
7. PIPE WRENCH
8. CHANNELLOCK PLIERS
9. ADJUSTABLE WRENCH
10. SOLDERING IRON OR GUN

FUSES AND CIRCUIT BREAKERS

Fuses and circuit breakers are designed to do exactly the same thing, but in different ways. Both almost instantly stop the flow of electricity in a circuit. Depending on their rating, they will stop the flow of current when it goes beyond a certain point. The idea, of course, is to stop the flow when too much electricity surges through, such as in the case of a short circuit, before damage can be done to the wires and units the wires are serving. If current continued to flow through a dead short, for example, the wires in that circuit would quickly overheat and a fire hazard would result.

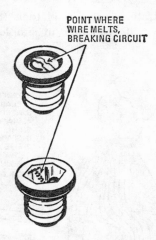

POINT WHERE WIRE MELTS, BREAKING CIRCUIT

The fuse works by melting the conducting wire inside. The wire is made to melt at a certain point and thus break the circuit and stop the flow of electricity. Some fuses will allow more current to flow before melting, others less. Each circuit in your home is rated to carry a certain amount of electricity, depending on the wiring which depends on the job of that circuit. If the circuits in your home are protected by fuses, each circuit will have (or should have) a fuse rated for that circuit. If something happens and a sudden overload develops (such as a short circuit or a dead short in an appliance), the fuse in-

GENERAL REPAIR PROCEDURES

stantly "burns out" (melts) and current flow stops. A burned fuse has done its job, and must be replaced after the cause of the burn-out is determined and corrected.

CIRCUIT BREAKERS IN A CIRCUIT BREAKER BOX

Circuit breakers do the same job, but in a way which does not require replacement if they are activated by an overload of current. An overload trips a spring and the breaker quickly opens, breaking the circuit and stopping the flow of current. A circuit breaker (you might find a small one on the back of your TV, and you will almost certainly find one on such appliances as garbage disposers, which tend to jam) can be reset after the problem which tripped it is eliminated. Let's say a short circuit occurs in an appliance, the breaker for that circuit trips, and the current flow stops. If the appliance is removed from the circuit, the breaker can be reset and everything else in that circuit once again has power.

Of course, when the fuse has been replaced or the circuit breaker reset, if the faulty appliance is plugged in again and the circuit shorted, the new fuse will melt or the breaker will trip again. Both are safety devices to protect the circuit. They will keep doing that job until the problem is solved.

4　　　　　　　　　SIMPLE APPLIANCE REPAIR

FUSE-TYPE BREAKER

Since circuit breakers are more modern and certainly more convenient than fuses and do the job just as efficiently and safely, there is now available a circuit breaker adapted for use in houses wired for fuses. This unit screws into the existing fuse socket, but acts in the manner of a circuit breaker. If it is tripped by a surge of current, it can be reset rather than needing replacement.

There is also available a "slow-blow" type fuse, which will accept a momentary surge of current before blowing. This might be used in a circuit where there is a motor which requires a heavy surge of current to get started, then trails off in current use to a standard amount while it is running. Formerly such a motor could blow the fuse protecting that circuit every time it tried to start. Of course even a slow-blow fuse will blow if the surge of current continues beyond a momentary, or safe, demand.

A serious mistake some people make is to defeat the purpose of a fuse by inserting an electrical conductor behind it in its socket. This is one of the disadvantages of fuses, especially considering they are used more in older homes where potentially more fire hazards might exist. If you know someone who might be tempted to defeat a fuse rather than locate the problem which blew it out, suggest using one of the tamperproof fuses now available. These look like regular fuses, but once they are screwed into their smaller sockets in a reducing adaptor in the fuse box, they cannot be removed without breaking them. Thus if a fuse blows, the temptation to remove it, place a penny behind it, and replace it, does not exist. If you remove it, you break it and must install a new one (which is what you should do in the first place). People who defeat the purpose of fuses in this manner are setting a time bomb in their homes.

It is almost as serious, but not quite, to merely replace fuses which are blowing with higher-rated fuses until the problem seems to have taken care of itself. The fuse is there to protect a circuit. It has been rated according to the circuit it protects. If you have added appliances and other electrical devices to an existing circuit until the fuse blows every time something turns on when other things are already on, the answer is not to increase the load rating of the fuse. The answer is to increase the load rating of the *circuit* by adding heavier wiring or, better yet, spread the appliances out over other circuits so that no one circuit carries the full load and each fuse can carry the load for which it was rated.

The same is true with circuit breakers, which are rated according to the circuits they serve just as fuses are rated.

Here's a trick to help locate a short circuit if a fused circuit continues to blow fuses. This should work if an appliance is drawing too much current. Turn off everything on that circuit. Remove the blown fuse and screw a hundred-watt light bulb in the fuse socket. If the bulb burns brightly, there is a short or too much current draw somewhere along that circuit. One by one, unplug all working appliances from that circuit. When the bulb goes off, the offender has been found. If everything is off and unplugged and still the bulb burns, the short circuit could be in the wiring. In that case, remove the bulb from the fuse socket and call an electrician.

TERMINALS, WIRE CONNECTORS, AND SPLICING

When you splice a wire, you will probably twist the two ends together and tape the joint. Such a connection will work for years if done correctly. But there are ways to connect and splice wires that will last forever.

In an appliance plug, for example, the wires are screwed in place. When a wire is attached to a plate, a screw is also used for permanency. Where a wire is held in place *only by a screw,* the wire should be wound clockwise around the screw to prevent its unwinding when the screw is tightened.

To be certain of the connection, a solderless terminal connector can be used. The connector is crimped to the wire and then a screw is used to connect the connector to the terminal.

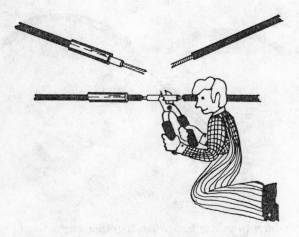

When splicing two wires together for a good mechanical and a good electrical bond, use of a splicing sleeve with a "spaghetti" insulator is recommended. The spaghetti is slipped over the wire and out of the way, then the bared ends of the wires are placed together. The splicing sleeve is located and crimped into the wires; then the spaghetti is slid down over the bond.

Another excellent way of splicing two wires together is with the use of a wire nut. Remove the insulation from the ends of the wires to be connected, twist the ends together clockwise, then press the wire nut onto the bond and turn it clockwise until it is securely attached.

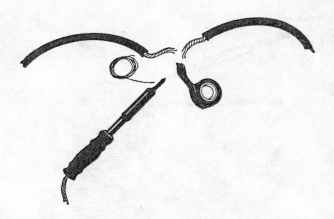

Here's how to solder two wires together correctly. (1) Strip away the insulation from each end. If needed, use some steel wool to brighten the area of the bond. (2) Clean the tip of the heated soldering iron or soldering gun with steel wool or a fine file. All surfaces of the iron tip should be shiny and clean. (3) Tin the soldering iron by applying a bit of solder to it. The solder should flow freely over the entire cleaned surface. (4) Twist the ends of the wires to be joined tightly together. This twisting will be the major source of strength for the bond. (5) Apply the hot iron to the joint and then apply some solder. As the joint heats, the solder should flow freely *into* the twisted wire bond (and not merely glob or cake on the surface of the bond). (6) After the joint has cooled (it should be bright and shiny, not dull and leadlike), apply insulating tape. The tape thickness, overlapped one way and then the other at least an inch past the bond, should be the same as the original insulation.

MALE AND FEMALE APPLIANCE PLUGS

If a problem is discovered in a male appliance plug it is quite easy to rewire the plug or obtain a new plug and wire it to the old appliance line cord. Whether the plug and line are three-wire or two-wire (illustrated), the procedure is the same. (1) Remove the insulator, then loosen the screws holding the

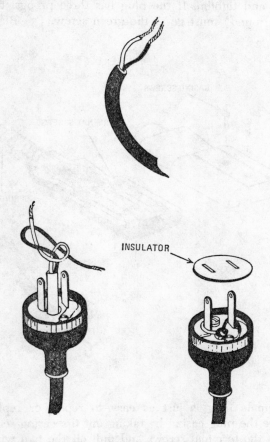

wires. (2) Remove the wires and cut or untie any knot between the two (this is a safety knot, to keep the wires from pulling out of the plug in use). (3) Pull the wires out through the hole in the plug, and insert them through the new plug (or, after stripping or removing any damaged portion, back through the old plug if it is to be reused). (4) Tie a knot in the two wires to prevent them from being pulled out through the hole in the plug, then reattach the wires around the screws and tighten. If the plug has three prongs, the green wire (ground) must go to the green screw. (5) Replace the insulator.

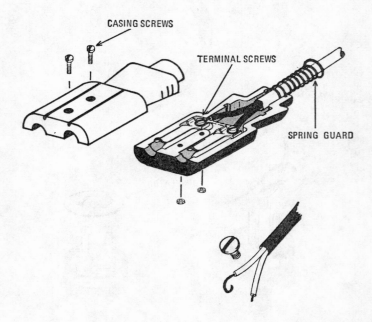

A female plug is just as easy to repair or replace. (1) Remove the plug casing by taking out the casing screws. (2) Loosen the terminal screws and pull off the two wires. Take them out through any spring guard present. (3) Insert the wires in a new plug, or cut off the damaged part of the wire and remove the insulation for reuse of the same plug. (4) Tighten the terminal screws. (5) Be sure the wire guard is caught between the case halves, then reinsert the case holding screws and tighten.

HEATING ELEMENTS

Many heating elements can be repaired if they are found to be broken (opening the circuit and allowing no current to flow). The connection must be physically good as well as electrically sound, and impervious to heat since the wire element will be heated to a glowing red. The best way is probably with a small brass nut and bolt, shown here with a wire heating element but equally possible with a ribbon element (though some repairmen prefer to rivet the wires together for a lasting bond).

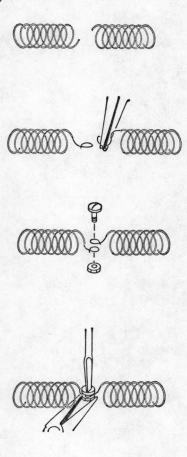

(1) Unplug the appliance from the wall, locate the break, then straighten the element ends. Make a loop in each. (2) Insert the bolt through the two loops. (3) Attach and tighten the bolt securely to complete the repair.

2

DISHWASHERS

When a modern automatic dishwasher is working properly, it is an economical, time- and work-saving machine. Using less water than it takes to do dishes by hand, less electricity than you would use to cook bacon and eggs for breakfast, and much less detergent than for hand washing, it washes dishes and pans sparkling clean—when it is working correctly. Not only are dishwashers responsible for a high percentage of unnecessary calls for outside service; they are also vulnerable to many exterior problems. The water may not be hot enough coming in. There may not be enough water. The detergent might not be fresh and strong, or the right type for your water and cleaning job.

In fact, there is a good chance that any given dishwasher is not doing the job as efficiently as it could.

HOW THEY WORK

Dishwashers work by a combination of hot water under pressure and the chemical action of soap. After you load dishes and pans and silverware into racks and latch the door, you push a button which starts the cycle. This activates the timer, which is the heart of the machine.

The timer, an electric clock that controls the different parts in the correct sequence, first opens the hot-water inlet valve and the machine fills to the proper level (controlled by a float and switch to prevent overfilling). Then the pump sends the hot water through spray arms as detergent is automatically

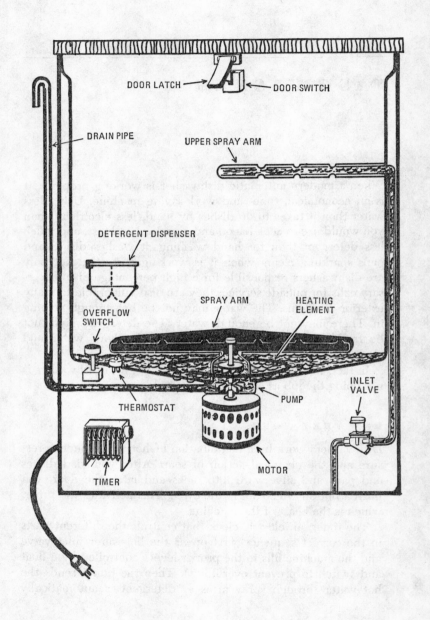

added from prefilled detergent cups. Different models go through different numbers of washes and rinses; then finally the pump removes the last filling of water through the drain pipe.

After the tub has been pumped dry, the timer turns on a heating element that has been maintaining the water heat. This time the heat (about 180° inside) dries the dishes as a vent opens to allow the hot, moisture-laden air to escape.

CHECKING THE DISHWASHER

After the tub has filled, allow it to empty and fill again, then press the latch and open the door. This will stop the action of the timer. Insert a thermometer in the water. A temperature below about 140° is not enough. Greases will not be dissolved and detergents cannot work properly. An ideal temperature is about 150°. Anything over 160° is wasteful and should be reset. If the water is not hot enough or is too hot, adjust your water heater (see Chapter 3).

Note also the level of the water in the tub. It should reach or cover the heating element at the bottom. If it doesn't, there may not be enough water to do a good job (see the section "Dishwasher Won't Fill" below).

Dishwasher detergent has a very short shelf life after the seal has been broken, sometimes as short as *two weeks*. Be certain the package you buy is sealed. Test the detergent on your shelf by dissolving a couple of teaspoonfuls in a glass of hot water (not less than 140°). If after two minutes a gritty residue is still evident, the detergent is not working as it should. Detergent which has caked in the box is generally worn out. Be sure to close the pouring spout on the detergent box to protect the contents.

The dishwasher will not work properly if it is not loaded according to the manufacturer's instructions, so look over the paperwork that came with the machine.

SIMPLE APPLIANCE REPAIR

DISHWASHER LEAKS

First of all, check to determine whether the leaking water is sudsy or clear. This will indicate whether it is coming from the inlet side of the machine or from the pump or drain hoses. Remove as many panels as you can to expose the hoses and parts. Tighten any hose clamps. If a hose is cracked, replace it, routing it the same way as the old hose.

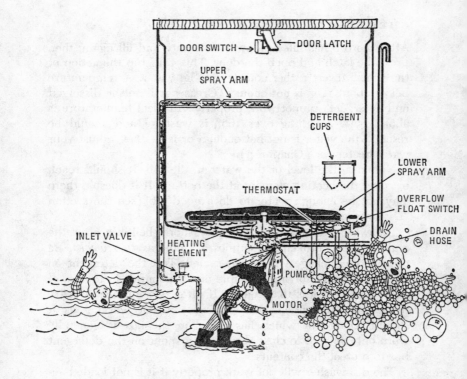

Inspect the area where the water enters the machine. Occasionally the inlet valve will leak, and sometimes a leak will develop around the pump housing. The valve can be

DISHWASHERS 17

replaced; and although the pump housing gasket is a bit more involved (first try epoxy cement around the area of the leak), it can be replaced by removing the screws holding the pump housing and installing a new gasket.

If the dishwasher is leaking because it is overfilling, the inlet valve is almost certainly at fault (or, of course, the float or switch). Unplug the machine (keeping out of the spilled water when you are handling the plug). Whatever else might be wrong, this should allow the inlet valve mechanism to close. If the water continues to flow, the inlet valve must be replaced.

If soapy, sudsy water appears around the drain hoses, check them for cracks or loose clamps. Tighten the clamps. Be sure to check where the drain hose attaches to the pump.

DISHWASHER WON'T FILL

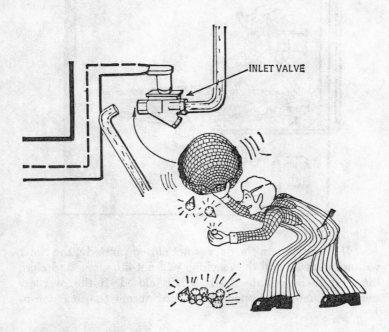

1. The inlet filter screen (or any filter screen on the faucet aerator of a portable unit) might be clogged with minerals. Clean the filter with a toothbrush and reinstall.

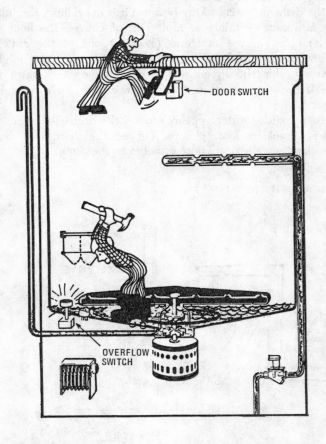

2. If the switch on the door is not closed properly, the timer will not activate and the machine will not fill. Open and close the door to be certain the switch is closed. If the overflow switch is jammed or open, it will not permit the machine to

fill. The job of this switch is to shut off the flow of water when a certain level is exceeded, but if it is not operating correctly, it will shut off the water too soon. Tap it. Both switches can be replaced by the appliance owner by following directions with the new switch, and referring to the material that came with the machine.

DISHWASHER WON'T EMPTY

1. The strainer of every dishwasher will become clogged sooner or later and need cleaning. It is the job of the strainer to protect the pump. It has openings large enough to allow soft food to enter and be washed away, but (hopefully) not large enough to allow chunks of bone, glass, or any other hard object to pass through and damage the pump. On some machines the pump housing must be removed to clean the strainer screen; on others simply removing a nut atop the lower spray arm will allow access to the screen.

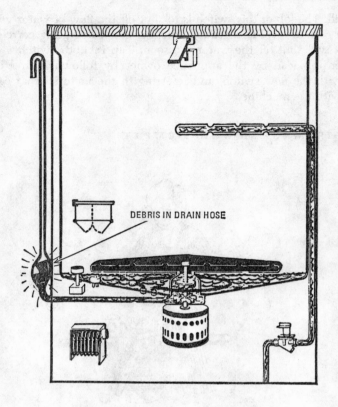

2. It is also possible that debris can get through the strainer and the pump, and clog in the drain hose. This will block the water and the machine will not empty. It is not difficult to remove the drain hose and blow it clean.

DISHWASHER WON'T CLEAN

1. The most common reason why a dishwasher won't clean is that the detergent is not correct, or is being used in an incor-

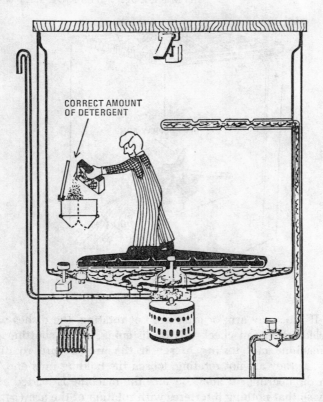

rect way. Refer to the manufacturer's instructions and recommendations. If these are not available, vary the amount of detergent up and down until the dishes are washed clean (remembering that too much detergent can get caked in the cups and never drop into the machine). If this doesn't work, try different types of detergent.

2. It is also quite possible that the water is not hot enough to do a real cleaning job. Refer to "Checking the Dishwasher" at the beginning of this chapter.

3. If the spray arm or arms are not rotating, the dishes will not clean. You can check this by listening, or by shutting off the machine and testing to see if the arm(s) are rotating freely. If they are not rotating, loosen the holding nuts and/or clean the bearing surfaces. Be sure the machine is loaded correctly so that nothing interferes with rotation of the arm(s).

DISHWASHER WON'T DRY

1. The machine must be given time to do the job, even when it is working properly. Yet many people rush to open the door and remove dishes immediately after the cycling has stopped. Open the door a crack and wait five minutes before removing dishes if they are coming out damp.

2. If the water in the dishwasher is too cool, the dishes will not dry properly (nor will they wash properly). Refer to "Checking the Dishwasher" at the beginning of this chapter.

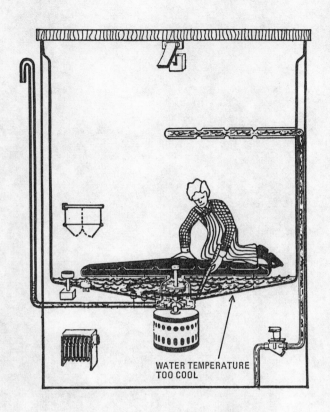

3

WATER HEATERS

Not that many years ago the water heater in most homes was a tank connected to the cooking stove in the kitchen, or sometimes to the heating stove. Water stored in this tank was heated whenever the range or furnace was in use. Such units were still in common use as late as the 1950s in some areas, offering hot water around dinner time and cold water the rest of the day.

Most homes today are equipped with automatic water heaters which use electricity or gas to heat the water. These heaters store hot water and dispense it as needed, heating a new supply as the water is used.

HOW THEY WORK

No matter how the water is heated, by gas or electricity, most water heaters work about the same way. Water enters the tank (either through the bottom or side, or through the top and to the bottom through a "dip tube") and is heated. A thermostat controls the source of heat, activating when the water cools or when too much cold water enters (because hot water is leaving). Hot water passes out through a pipe at the top of the tank and to wherever the open hot-water faucet is in the home.

Tanks have (or *should* have) a pressure relief valve at the top which, if activated by too much pressure, vents off hot water and pressure as a safety measure. Modern tanks also

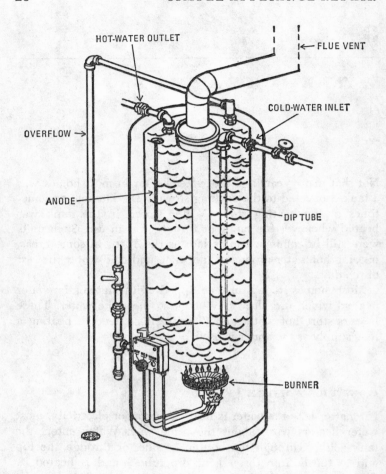

have a "sacrificial anode" suspended inside. Because of the action of the chemicals in the water against any metal inside the tank, corrosion and eventual failure could result. The anode, made of more readily corroding magnesium, is there to corrode, thus saving the metal of the tank. Also, most modern water-heater tanks are glass lined (really a plastic) to cut down on water-to-metal contact and subsequent corrosion.

WATER HEATERS

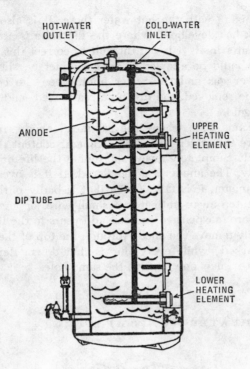

Tanks are wrapped in a blanket of fiber glass insulation (between two walls) to help contain the heat and save energy.

CHECKING THE WATER HEATER

It is possible that the dip tube can come loose inside a hot-water tank, allowing cold water into the top of the tank rather than directing it to the bottom. This is called "bypassing."

If your tank is providing a large amount of warm water but no hot water at all (regardless of the thermostat setting), it is likely that the dip tube has broken loose at the top and cold water is mixing with the hot water there. It is also possible that a leak has developed where the dip tube is attached to the cold-water inlet. It is not difficult to correct this problem. The tank should be allowed to cool completely; shut off the electricity or gas and the water inlet. Then the cold-water inlet can be removed, the dip tube reattached, and the inlet fastened again.

At the same time, check the sacrificial anode. In certain areas where the water has a high mineral content the anode may corrode completely away, shortening the life of the tank significantly. The anode is a metal rod. If it is over half an inch in diameter, it is O.K. (even if it is badly pitted). If it has been eaten away until it is a small wire, or less, replace it. Anodes are inexpensive and can add years to the life of the tank. Merely remove the anode through the top of the tank (it will be marked) while everything is shut down. Replace the old one with a new one through the same hole.

WATER HEATER WON'T HEAT

1. If the water won't heat in an electric heater, the problem is generally that no electricity is reaching the heating elements. Check for a blown fuse or open circuit breaker and be sure the thermostat is set high enough so that the elements will energize with electricity.

WATER HEATERS

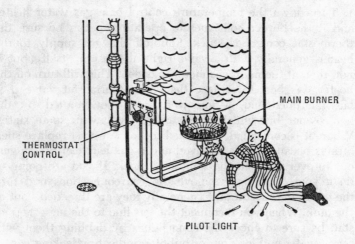

2. In a gas water heater, check the pilot light to be sure it is on. If the pilot goes out, the unit has a safety feature which will shut off valves and allow no gas to flow. Before you relight the pilot (following the instructions on the unit) be sure there is no lint or other flammable material around the orifice and that all baffles are in place.

WATER TOO HOT

1. Turn down the temperature control of a gas water heater and if the burner continues to operate you can be sure the thermostat control is stuck. Shut off the gas supply to the heater, generally with a valve right at the unit itself (but if not, then at the main gas meter outside). This will turn off the heater. It's best to replace the entire thermostat valve if it has malfunctioned, a job which can be handled by the homeowner but which is controlled by law in some states. Where it is controlled, a licensed technician must replace such fittings because gas is involved and a gas leak can be dangerous to everybody in the same city block. If you are going to do it yourself, simply remove the gas connections and turn the entire thermostat unit (generally they are threaded) out of the tank. When you reconnect the gas line to the new thermostat, be sure to check the connections by dabbing them with soapy water and looking for bubbles, indicating leaking gas.

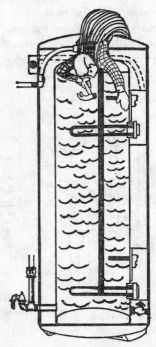

WATER HEATERS 31

2. In an electric water heater, you may have an improperly adjusted thermostat; simply turn it down. More likely, the double-throw thermostat (there are two heating elements in an electric water heater) may be sticking in the closed position, causing the upper element to operate all the time. The top of the tank will become dangerously hot, while the bottom three quarters of the tank will stay cool. The thermostat must be replaced.

NOT ENOUGH HOT WATER

1. If there is not enough hot water, the remedy is usually a simple matter of increasing the thermostat setting on either gas or electric water heater. But also refer to the discussion of bypassing in "Checking the Water Heater." Remember, about 160° is *plenty* hot in a tank.

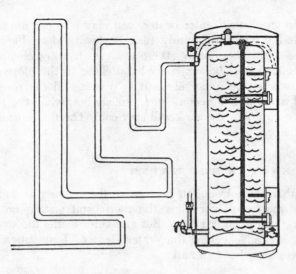

2. It is possible that the heater, regardless of type, is doing the best job it can under the circumstance of too extensive a pipe system. Many large homes have two or more water heaters, or two heaters connected to a large central storage tank. If the hot water must travel a great distance to reach the faucet, it is possible that it will cool down some.

LEAKING

1. Although it is possible that the bottom of the tank is leaking through a corrosion-caused hole—and probably sooner or later this will happen—generally leaks can be traced to where the pipes enter the tank. Tighten the connections if you see a drip. Unscrew them and use a pipe-thread sealing compound if tightening doesn't stop the drip.

2. If too much pressure builds up inside the tank and there is no overflow, or if the overflow is stuck, leaking can occur. If too much pressure is present in the house lines, leaks can be forced in the fittings and connections.

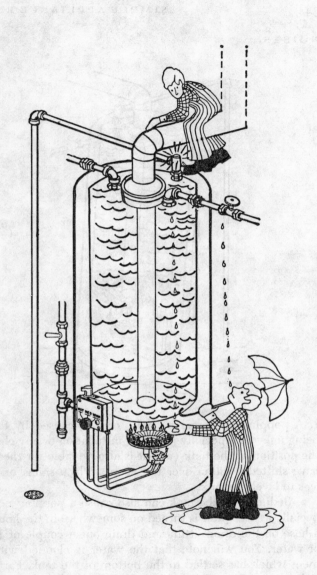

NOISES

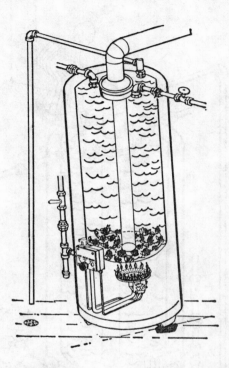

1. A popping and cracking noise can be caused by the tank being uneven, especially just after installation or any change in the position of the tank (and it is also possible for the unit to have shifted position over the years). Place shims under the legs to level the tank.

2. Sediment in the tank can also cause a popping noise, especially as the water is turned on somewhere in the house. Put a hose on the drain fitting and drain out a couple of buckets of water. You will note that the water is clouded with sediment which has settled to the bottom of the tank. Part of the sediment, in fact, is probably bits of the sacrificial anode.

WATER HEATERS

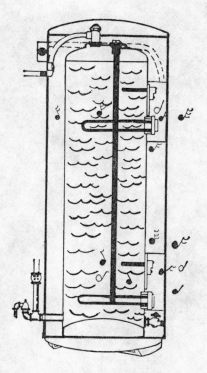

3. Earlier heating elements in electric water heaters were noisy due to rapidly boiling water surrounding their sheath. This often sounds like a "singing" noise. If your electric water heater is a few years old and it "sings" to you, you probably have a high-watt-density heating element. Don't change the element just because of the noise; the sounds are there because the element is doing its job efficiently, and no harm will be done to the tank, the element, or the water. Live with it until the element gives out, then replace with the newer low-watt-density elements. These spread the heat input over a larger area and eliminate the sounds.

4

WASHING MACHINES

It does seem as though you can dump in a load of dirty clothes and the new machines will wash, soften, sweeten, dewrinkle, fluff, and almost fold them for you. Clothes washing machines have cycles for nearly everything. Almost three quarters of all service calls on automatic washers today are for very simple repairs—belt replacing, switch adjusting, and so forth. These problems, as well as a more complicated water-pump change, can be handled by the householder.

HOW THEY WORK

The operating principle of a modern automatic washer is the forcing of hot, soapy water through the fabrics by mechanical action (usually agitation) to clean them. A timer, a built-in electric clock, is turned on to begin the actions inside the washer. First, a mixing valve is opened and the tub fills with hot, warm, or cool water—the exact temperature depending on which of the several cycles of the timer has been selected (woolens, cottons, wash and wear, etc.). The water level is sensed and when the proper amount of water is in the tub, again according to the cycle selected on the timer, a pressure switch stops the flow. The timer also starts and stops the motor that works the agitator, the pump that empties the tank between washes and rinses, and the spinning of the basket in the tub to begin the drying of the load.

On many modern machines the pump also circulates the water in the tub through a lint filter to keep the wash water as clean as possible. Then, as the tub is emptied, a reverse flow is set up through the filter, washing the lint into the drain pipe and cleaning the filter for the next wash cycle.

If the load shifts to one side of the basket, an off-balance switch is tripped and the motor shuts off. On most machines a buzzer also sounds to inform the operator of the off-balance condition. Still another switch is built into the lid of the tub, automatically shutting off the spinning of the tub if the lid is opened during that cycle.

CHECKING THE WASHER

If the automatic washing machine is not cleaning properly, check the temperature and pressure of the water coming in. Unlike a dishwasher, which has a heating element inside, clothes washers depend on water heated somewhere else and piped into the machine. The washer assumes that correctly heated water will be available at one inlet and cold water at the other, then mixes them according to the cycle you have selected.

If the pressure of the water is not high enough, water will not enter or mix properly. (Too *much* pressure can damage the valves inside the washer; so manufacturers often suggest turning off the water when the machine is not in use.)

The water from the hot-water faucet should be no less than 140° and no more than 160°. Adjust the water-heater thermostat to bring the water within this range. Water pressure should not exceed fifty or sixty pounds if you hope for long life from home appliances using water. You can buy an inexpensive pressure testing gauge which screws onto the faucet just like a garden hose; when you turn on the water, the gauge indicates the pressure. If the pressure is much too high or much too low (and the pressure from the feed lines into the house is high enough), suspect any pressure-regulating

WASHING MACHINES 39

device between the incoming water line and the house main supply (on the house side of the water meter).

Normally if the pressure-regulating valve has failed, it will allow full pressure from the city lines into the house lines. Repair is simple though, and a repair kit can be obtained from most home do-it-yourself stores and plumbing stores. Dismantle the valve, replace the old diaphragm and springs with the new parts, and reassemble.

MACHINE WILL NOT FILL

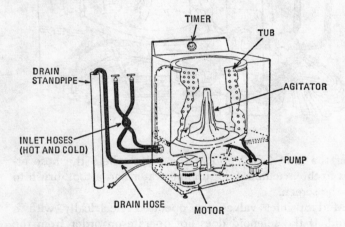

1. If you are sure the faucet is turned on, but the machine does not fill, the problem could be something as simple as a crimped or kinked inlet hose. This might happen if the washer has been moved. Unfasten the hose from the faucet, straighten it, and reconnect.

2. Washers have filter screens in the inlet valves to stop sediment and minerals from entering the system from the supply lines. Sand, scale, or rust particles may have clogged the screen and impeded or stopped the flow of water. Remove and clean the screen or screens at the coupling (the faucet

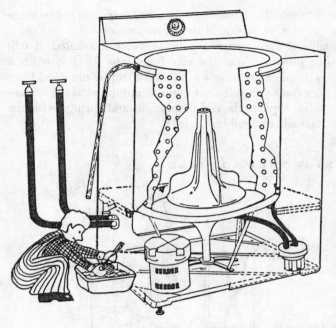

end of the inlet hose) and at the valve where the hose attaches to the machine. Use running water and a toothbrush to clean the screen.

3. Modern inlet valves are operated electrically with a solenoid. If the solenoid does not operate on order from the timer, the valve will remain closed and no water will enter the machine. Each inlet valve will have a solenoid, which you can find by lifting the top or removing the back panel on most washers. If you hear a slight humming or buzzing, the solenoid is probably working (an electric current creates a magnetic field which holds the solenoid open during operation). A tap might free it if it is stuck closed, or, if that doesn't work, a new solenoid can be mounted. Replace the unit with another exactly the same (always unplug the machine before any testing or repair work), attach the two wires, and reinstall the top or back panel.

WASHING MACHINES

TOO MUCH WATER IN THE MACHINE

1. Unplug the machine. If the water shuts off, the water-level switch is probably at fault. Be sure the tube allowing water to the switch is not clogged. If the switch still doesn't work, replace it.

2. If the water continues to enter the machine after it is unplugged, it is possible one of the water inlet valves is clogged and not closing completely. A small grain of sand or debris from incoming water might have bypassed the filter screen and lodged in the pressure balance passageway of the valve. It is not a difficult job to remove the inlet valve, disassemble and clean it (be sure the filter screen is completely clean, and checked carefully since the debris should not have passed), then reassemble the valve. This is a worthwhile procedure even if it takes a little time, since a serviceman would probably just replace it and charge you for a new part plus an expensive service call.

MOTOR WON'T RUN

If the motor won't run, probably the trouble is at the plug in the wall and not in the machine. Check the voltage at the plug with a lamp (see the instructions in the Foreword).

MACHINE WON'T RUN BUT MOTOR HUMS OR RUNS

1. Remove the back panel of the machine or the access panel to the motor. An automatic washing machine is driven by one motor which turns belts or gears to run the tub, agitator, etc. If the motor is running but not turning the main drive belt or belts, check to be sure the belt is tight enough and not broken. Some belts can be replaced easily by sliding them over their pulleys. Others can be replaced only after certain other components have been removed. A loose belt is generally tightened by loosening the motor mounts and then moving the motor slightly to tighten the belt. A properly adjusted drive belt can be deflected about half an inch midway between the pulleys with moderate thumb pressure.

2. If the motor is humming but not running, check the drive belts to be sure nothing is binding (watch out for your fingers . . . shut off the machine). If you can flip the main belt off the motor, you can see if the motor will run alone. If it doesn't, remove the motor and have it checked. Sometimes centrifugal switches or capacitors have failed and can be

WASHING MACHINES

replaced by a motor repair shop. Be sure that the problem was not simply a matter of too large a load—or too much detergent—interfering with the operation of the machine.

AGITATOR WON'T OPERATE

1. Check the belts as discussed in the preceding section. An adjustment of the drive belt should correct the problem unless the belt is broken. Replace the belt if it is cracked, slipping, or broken.
2. The clutch or timer may be at fault. First check for loose wires on the timer, after unplugging the machine. If the timer needs to be replaced, or the clutch adjusted or replaced, the job is a little more difficult. You can do it by carefully following the parts and service books that came with the machine, but undertake it only if you are comfortable with it. Otherwise, call in a serviceman.

LINT FILTER NOT WORKING

1. You should be aware that certain items of clothing should not be washed with certain other items. For example, items that tend to collect lint should not be washed with items that tend to give off lint, otherwise the one will merely hold on to the lint from the other and the lint filter will appear to have failed. If your machine lint filter seems to be falling down on the job, try a few loads of carefully separated clothing, and add a good fabric softener to be sure.
2. It is also possible that a small article of clothing has entered the piping to the filter and blocked it. On some machines you can see the action of the filter by watching the filtered water flowing back into the machine. On motor reversal types you must check the components to see if they are clogged.

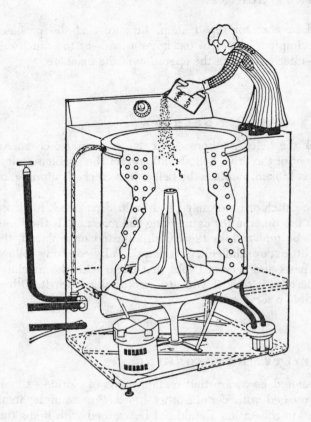

MACHINE LEAKS

Automatic washing machines generally leak at the inlet hose connections or the drain hose connections, seldom in the tub. Heat expands fittings and softens hoses. Trace down the leak and tighten the connection or replace any damaged hose. If a leak is coming from the pump or some other housing or body, replace that housing (but be sure it isn't just a matter of replacing a gasket between the main tub and the housing).

WASHING MACHINES

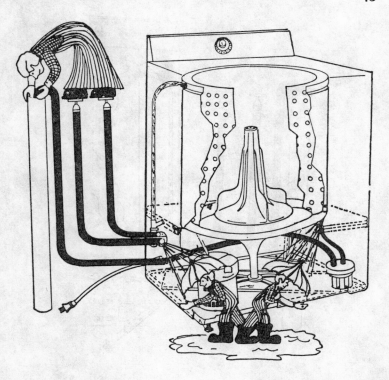

Locate exactly where the leak is coming from. If the leak is coming from the drain standpipe, check to be sure the pipe is not clogged with lint or debris.

MACHINE WON'T EMPTY

1. If the machine won't empty properly, most likely the drain hose or drain standpipe is clogged. Remove the hose from the pump to the standpipe and clean it out. Check the standpipe or other drain outlet.

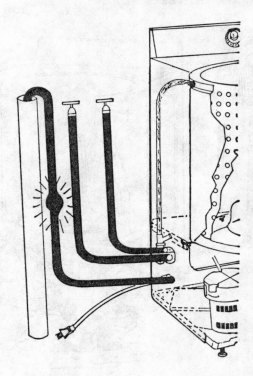

2. It is also possible that a small article of clothing has been sucked into the drain outlet of the tub, or that the lint filter or even the pump is malfunctioning. Check the tub outlet, and check to see if water is being pumped from the outlet side of the pump by watching for water from a cleaned drain hose. Refer also to paragraph 2 in the section "Lint Filter Not Working."

EXCESSIVE VIBRATION

1. The most common cause for excessive vibration in a clothes washer, and the thing that will eventually shut off the ma-

WASHING MACHINES

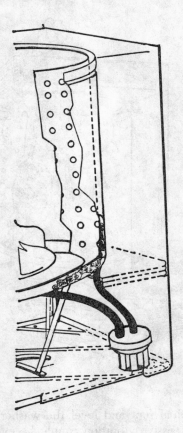

chine, is an unbalanced load in the tub. Although modern washers are designed to absorb some unbalance shock without transferring the vibration to the body of the unit, severe unbalance will stop the operation. Simply rearrange the clothes in the basket. Watch for a towel or other item that's extra heavy with absorbed water and spread it around the agitator.

2. Most washers have levelers at the bottom—flat, broadheaded screws which can be turned in or out to extend or re-

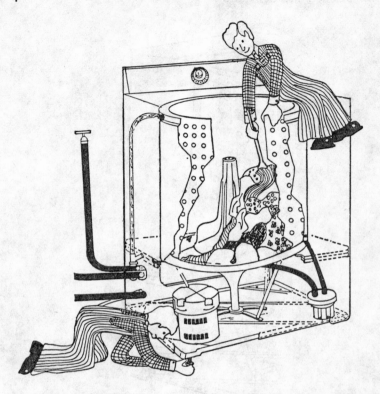

tract each individual foot and level the washer on the floor. Check these if excessive vibration cannot be explained. Place a level on the top of the washer and adjust the feet until level is perfect. The washer should rest on a solid floor, or vibration can be transferred and amplified.

5

CLOTHES DRYERS

Gas or electricity is the source of heat to dry clothing in a modern automatic clothes dryer. This heat is controlled by one or more thermostats of bimetallic construction. Clothes are put into a drum which is turned by a motor-driven belt while heat is blown through by a fan. At the end of the drying cycle, when the water has been removed and the heat rises inside the drum, the gas or electricity is shut off by the thermostats and the job is done.

The whole thing is controlled by a built-in timer similar to an electric clock. Most dryers also have a lint filter that collects bits of debris from the drying clothes.

HOW THEY WORK

When you set the timer to the operation desired, a motor turns on and the drum holding the clothing and a fan are activated. At the same time, the heat source turns on. In an electric dryer, current flows through heating coils; in a gas dryer, a pilot or an electric coil turns on a gas flame in a burner. The exhaust fan draws the warm air through the tumbling clothing.

Thermostats monitor the temperature of the air in the exhaust duct, and a trap in the same duct collects lint and prevents it from blocking the flow of air.

If the temperature in the duct rises (as when the clothes are dry) or when the timer orders it, the machine shuts down.

CHECKING THE CLOTHES DRYER

Modern dryers are relatively simple mechanisms, although a gas dryer is somewhat more complicated than an electric one due to the extra gas piping involved. They either dry the clothes load or they do not. If the unit runs, if sufficient temperature is available to dry the clothes, and if the unit shuts off when the job is done, it is working correctly. Temperatures and drying times vary with venting (or room size), with gas or electric supply, and with size of load to be dried.

DRYER WON'T RUN

1. Check the fuse box or circuit breaker box for a blown fuse or open breaker. Check the plug in the wall for current by unplugging the dryer and plugging in an electric lamp.

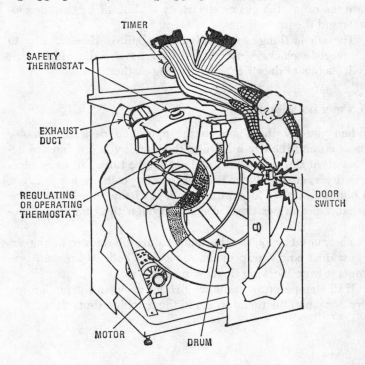

CLOTHES DRYERS   51

2. Most modern dryers have a door switch that prevents operation when the door is open. If the door is not shut completely or properly, the unit will not start. Check the door switch by opening and reclosing the door firmly. Although it seldom fails, the door switch can be replaced easily if failure should occur.

DRYER WON'T HEAT

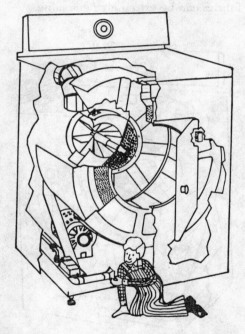

1. Electric dryers and some gas dryers depend on electricity for heat. In the electric dryer, all of the heat is provided by heating coils energized electrically. In many gas dryers, electricity energizes a glow coil when the timer is activated and the glow coil acts as a pilot light for starting the main gas supply in the burner box. Thus when the dryer is shut off be-

tween uses, there is no gas flow (as for a constantly burning pilot light) and no current flow.

2. Be sure the gas supply is turned on, and that the burner is not clogged with lint or debris. Too much lint anywhere around the pilot circuit or the coils or burners will prevent the unit from heating. Many servicemen recommend that you remove the access panels once each year and vacuum out the accumulation of lint and dust. Not only will such debris prevent proper operation, but the residue of lint and debris left by modern fabrics can be extremely flammable.

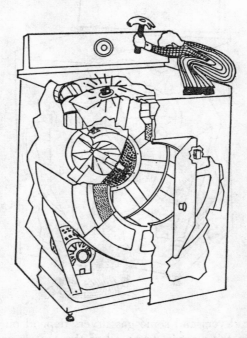

3. Modern dryers are equipped with a safety thermostat which prevents overheating. If this thermostat fails, the dryer will not operate. This thermostat is normally located on the housing containing the heating elements or the burner box. The thermostat will open, breaking the circuit to the heat

CLOTHES DRYERS

source, for a number of reasons, including fan-belt failure and blockage or impeding of the air flow.

CLOTHES DO NOT DRY

1. It is possible that your automatic washer may be affecting the operation of your dryer. If the clothes load is too damp when it is put into the dryer, it may not dry properly. Check the operation of the spin cycle of your washer.

2. An inadequate air flow through the clothes load will also impede the drying process, and the most likely place for the air to be slowed is where lint is collected from it or at the exhaust point. Check for blockage or lint. Be sure any vent tube or device on the outside of the wall is clear. Check to be sure

the fan is operating. A broken fan drive belt is easy to replace by removing the rear access panel.

TOO MUCH HEAT

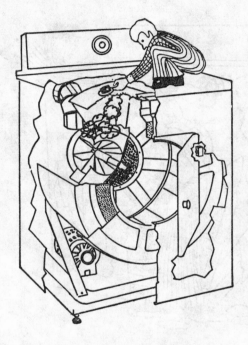

Electric and gas dryers are equipped with regulating thermostats to prevent too much heat. If the heat rises too high, the thermostats cut it off until the temperature drops. But if the regulating thermostat becomes covered and insulated with lint it cannot sense a rise in temperature. Remove any lint from the vicinity of the regulating thermostat.

CLOTHES DRYERS 55

DRYER CONTINUES TO RUN WITH DOOR OPEN

The door safety switch is there to prevent the machine from operating with the door open. If the dryer continues to operate, not only will clothing be dumped out on the floor in some models, but it is also possible for a person to be caught up in the turning drum. If the safety switch has failed, it should be replaced as soon as possible, especially if children might be in the vicinity of the machine. The failure of this switch will normally not affect the operation of the machine, but it is a safety hazard.

NOISY OPERATION

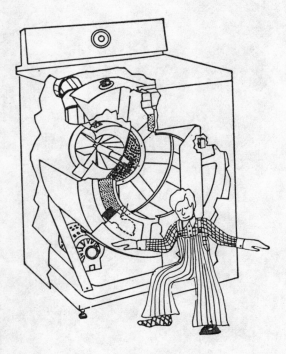

1. Check the level of the dryer by placing a carpenter's level on the top; if the unit is not level, adjust it with the leveling screws at each foot.

2. A loose fan or motor pulley or a loose or dried-out drive belt can cause noisy operation. An intermittent or regular squeak may be caused by a dry belt; it can be stopped by an application of belt dressing (available at any service station or appliance store). Loose cabinet screws rattling about can also cause noisy operation.

3. Any debris which has worked its way between the drum and the drum housing will also cause noisy operation. Turn the drum manually (with the dryer off and unplugged) to determine if something is caught between.

6

REFRIGERATORS AND FREEZERS

One family recently bought an older home and found in the kitchen wall a pass-through to the outside with heavy insulated doors inside and out. Both doors opened to a small compartment which adjoined a larger, cabinetlike shelved section. Can you guess what it was? If you can, you've been around for a while. It was an icebox, with provision for delivery by the iceman. He could drop off a fifty-pound chunk through the outside opening directly into the cooler. The housewife had access to the ice through the inside opening.

Little could go wrong with an icebox, but the convenience of a modern refrigerator and freezer far outweighs the few problems they can hand you. The machinery inside these two appliances is noted for being some of the most long-lasting of any in the home.

Modern refrigerators are cooled by their freezing compartments, while freezers are one huge freezing compartment.

HOW THEY WORK

Freezers work on the principle that a liquid absorbs heat as it vaporizes. If a liquid is vaporized under controlled conditions, as it passes through an area where you wish to remove heat, you could cool that area. That is what happens in a refrigerator freezing compartment or a freezer. And it happens in a "closed system" so that the vaporized liquid can be condensed back into a liquid form, the heat removed, and then go through the cycle again and again.

The liquid, called a "refrigerant," is pumped by a compressor through a condenser under high pressure. As it flows, giving up its heat to the outside air (you can feel this heat coming from your own refrigerator or freezer), it comes to a restriction like a capillary tube. This acts to reduce the pressure in the system beyond the restriction, and as the liquid enters the lower-pressure area (the evaporator, or "freezing unit"), it vaporizes. As it does this, it collects heat from inside the appliance and the food stored there. But it is still being moved by the compressor, so it passes out of the evaporator and back into the condenser, where it once again becomes a liquid and gives up its heat. A fan moves air across or through the condenser coils, blowing away the heat as the gas condenses into a liquid.

This cycle is continuous as long as the refrigerator is plugged in and turned on. It is controlled by a thermostat set by you. When the desired temperature is reached inside the box, the thermostat opens and shuts the compressor off. Everything stops. Then when the temperature rises above the point selected, the thermostat closes and the cycling of the refrigerant begins again.

Modern refrigerators and freezers have automatic defrost systems that prevent a build-up of frost on the evaporator unit. This is sometimes done by a timer which merely shuts off the system for a period of time to allow the frost to melt no matter what the thermostat says, or more often by a heating coil which heats up at preset intervals just long enough to melt the frost (which is caught in a runoff pan and evaporated as the heat from the system passes around it).

A modern refrigerator has a life expectancy of more than fifteen years, and with proper care and maintenance this span can be increased substantially. Can you guess how many homes now have refrigerators? Nearly every single one. Of all the homes in the United States that have electricity, 99.9 per cent have one or more refrigerators.

REFRIGERATORS AND FREEZERS

CHECKING THE CONTROLS

Most modern refrigerators have two controls, one to regulate the temperature at which the thermostat turns on and off, and the other to regulate a vent that allows more or less cold air from the freezer compartment to the rest of the box. A freezer generally has only the thermostat control.

Check the temperature inside the freezer compartment with a thermometer. It should be about 5°. The temperature in the refrigerator compartment should be about 35°. Be sure the air ducts are not obstructed and the food is stored in such a way that it allows a free flow of air throughout the unit.

TOO MUCH FROST

1. One of the most overlooked items in a modern refrigerator is the evaporator coil, especially since it is assumed to be working if it is frost-covered. But too much frost can cut the efficiency of the system, and this is often the direct result of foreign matter on the surface of the coil. The trouble is, if you touch it to clean it, you are leaving an oily residue which will soon collect water droplets and more frost. The coil also takes up a natural residue from the surfaces of items stored in the freezer.

The best way to start is to wash the evaporator coil with a dishwasher detergent mixed in a mild solution of warm water. Be sure to use a detergent for dishwashers, since other soaps can make the problem worse. A dishwasher detergent (use one marked "for use on aluminum") breaks down the surface tension of the water.

2. If the door seal is leaking warm air into the box, too much frost can result. An easy way to check to see if the door gasket is sealing is the "dollar-bill test." Close the door on a dollar bill at several locations around the seal, then pull the bill out. There should be a definite resistance as the bill slides out.

DOLLAR-BILL TEST

If you find an area where the bill slides out without resistance, the door seal is probably leaking. Loosen the screws which hold the inner door panel then gently twist the door (slightly warping it to realign the seal) until proper alignment is achieved. Then retighten the screws and make the dollar-bill test again.

WON'T RUN

1. If your refrigerator or freezer won't run, check the plug in the wall to be certain electricity is available (if it's not, check the fuses or circuit breakers) and check the plug from the

REFRIGERATORS AND FREEZERS 61

appliance itself. These are larger plugs and occasionally the wires loosen where they attach to the plug.

2. Refrigerators and freezers have a thermostat switch which also acts as an off-on switch. Turning the thermostat switch too far shuts off the appliance; so if it won't run, be sure the switch is on.

TOO COLD

1. The most likely cause of a refrigerator or freezer being too cold is that the thermostat switch is set too low. The temperature of the room in which the appliance is being used has a

REFRIGERATORS AND FREEZERS

definite effect on the temperature inside the box. If the room is normally cool or cold, the thermostat control can be set higher since too low a setting can freeze food inside the refrigerator part of the appliance. At high altitudes the lower barometric pressure will lower the range of the control.

2. It is possible for the thermostat control to malfunction and keep the unit running too much, or constantly, and the result will be too low a temperature inside. To replace the thermostat control, unplug the power cord, then unscrew the thermostat and gently pull it out. Unhook the two wires to the control, then hook them to the new control. Attach the new control the same way the old one was attached.

NOT COLD ENOUGH

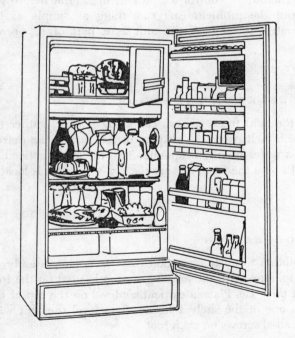

1. Too much food stored inside, or improper storage of food, can affect the temperature of the unit by blocking the flow of air, or even slowing it or changing its pattern. Before you do anything else, be sure the box is not overloaded or that certain items are not blocking the air circulation inside.

2. Leaking gaskets in the door can raise the temperature several degrees inside (and cause the appliance to run too much). Try the dollar-bill test described under "Too Much Frost."

3. The thermostat might be malfunctioning, not allowing the unit to run long enough. Unplug the power cord, unscrew the thermostat, and gently remove it. Then disconnect the two wires from the thermostat and tape them together with electrical tape. Plug in the power cord. If the unit operates normally, cooling down the box to the correct temperature, replace the thermostat control with a new one. (But before you do, be sure the problem isn't something as simple as the defrost control turning off the unit as a part of its normal cycle.)

WATER LEAKS ON THE FLOOR

Refrigerators and freezers have collecting pans to gather the water which melts from the evaporator coil during the defrost cycle. This water usually evaporates into the air unnoticed. If water leaks onto the floor, be sure the collecting pan is located correctly.

NOISY OPERATION

1. The solution to noisy operation may be as simple as leveling the appliance with the leveling screws located on the four corners at the base. Place a carpenter's level on the top of the box or on one of the shelves inside, then level the unit with the flat-headed screws on each foot.

REFRIGERATORS AND FREEZERS 65

2. Noisy operation may be as complicated a problem as a frozen divider block in the compressor. If the noise seems to come from the compressor, unplug the appliance and call a serviceman. You can replace a compressor yourself, but since modern refrigerators and freezers are hermetically sealed units, only servicemen have the equipment to reseal and pressurize the system after replacement.

SIMPLE APPLIANCE REPAIR

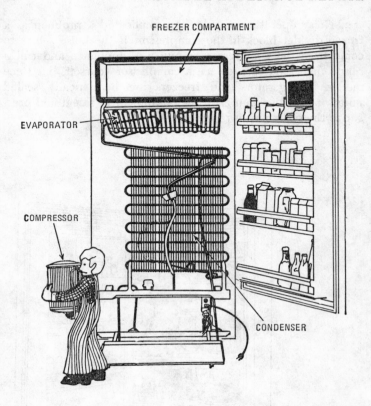

EXCESSIVE POWER CONSUMPTION

1. The condenser coils are located where you can get at them, either at the bottom of the refrigerator or freezer behind a snap-out panel or at the back. If you haven't seen the ones in your unit in a while, take a look. Chances are they are loaded with lint and dust wherever they are located. This coating reduces the efficiency by preventing a free flow of air around and through the unit. In many units, dust- and lint-laden air is

REFRIGERATORS AND FREEZERS 67

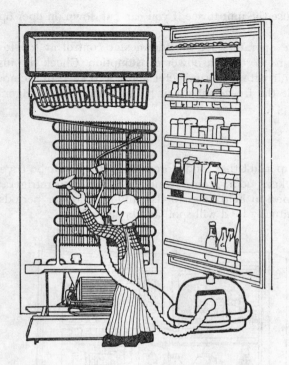

forced around the fins of the coil by a fan, so the problem is even greater. The condenser coils should be vacuumed for efficiency and economy at least once every six months.

2. Air leaking into the unit around a malfunctioning door gasket can cost a substantial amount of money as the box tries to overcome this warm air by running too long and hard. Try the dollar-bill test described under "Too Much Frost."

3. Every time you open the door on a refrigerator or freezer, you allow cold air out and warm air in. Check the

way you use the unit to see if you can cut down on door openings.

4. Some operators set the thermostat control at too low a temperature, costing in power consumption. Check the inside of your unit with a thermometer and try to keep the temperature inside the freezer at about 5° and inside the refrigerator at about 35°.

ODORS

1. It is important that food stored in a refrigerator be covered tightly to keep odors from mixing. Also, although refrigerators are designed to help keep foods edible for longer periods of time, eventually food will spoil and create odors.

REFRIGERATORS AND FREEZERS

2. After a period of use, refrigerators and even freezers tend to pick up odors. You may not notice them individually but finally the whole box begins to smell. Check and clean the drain pan, where bits of food can end up. Clean the interior of the box with warm water and baking soda, a teaspoonful for each quart of water. Be certain you unplug the refrigerator or freezer before cleaning and do not clean around electrical units with water or a wet cloth.

7

TOASTERS

Modern toasters are generally welded and riveted together inside. They are not made for the home handyman to repair when a problem occurs. Nor, as a matter of fact, are they made for a professional serviceman to repair. Toasters are designed to last for years, toasting bread morning after morning, then be replaced when they finally break down.

But there are certain things that you can take care of.

HOW THEY WORK

Modern toasters are designed to brown bread as conveniently as you are willing to pay for. Some toasters lower the bread automatically; most have a switch which you must press down to activate the toaster. Then, when the toast is done according to a dial you have set, a spring or a dash pot, a vacuum device, pops the toast out of the machine.

Inside the toaster, the toast is browned by a heating element of flat nichrome wires wrapped around a mica panel. Setting the control knob to determine the brownness of the toast sets the distance a bimetallic strip must move before it makes contact with a solenoid which kicks the toast up away from the wires and shuts off the electricity. The browner you order, the farther the strip must move, and thus the longer the toast is next to the heating elements.

CLEANING

If your toaster begins to smoke, you know it is ready for a good cleaning. But it is better to clean the toaster on a regular schedule so that bread crumbs, butter, spices, margarine, and sugar don't accumulate and begin to cause problems. Unplug the toaster. With a flashlight, examine down inside each slot. Locate where bits of food have collected on the wires which separate and hold the bread from the heating elements and on the elements themselves. Insert a soft paintbrush into each loading slot and gently brush the debris down into the crumb tray. Do not turn the toaster upside down to shake out the debris. This will allow crumbs into contact with the heating elements, where they will stick (and possibly cause the elements to burn out at "hot spots").

To get rid of stubborn debris, you can wash the insides with a solution of household ammonia. Allow the toaster to dry for at least an hour before use.

Once the debris has been brushed into the crumb tray or washed away, remove the tray and clean it.

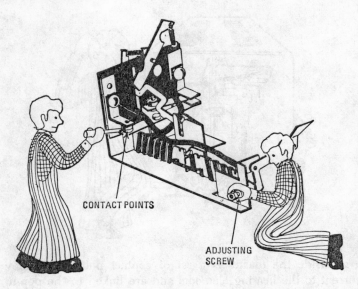

TOAST DOES NOT BROWN / TOAST IS TOO BROWN

1. The toaster will have a screw to adjust the timing mechanism so that the bimetallic strip will react the way you want it to. Perhaps the toast simply is not remaining down long enough, or is remaining down too long. You can adjust the screw by turning it in, or clockwise, to *decrease* the toasting time and out, or counterclockwise, to *increase* the toasting time. This screw, located in different places on different models, is in addition to the light-dark adjustment lever or knob on the toaster and turning it a bit won't damage anything. Be sure the toaster is unplugged when you make this adjustment.

SIMPLE APPLIANCE REPAIR

2. Inside the toaster is a set of contact points that allow current to the heating elements and are linked to the pop-up unit to turn the heat off when the toast rises. It is possible for these points sometimes to stick together (the elements will not shut off and the toast will get too brown) or not to make contact at all (toast will not brown because the elements will not turn on). If the points tend to stick together, or to arc (electricity jumps across between the points), you should clean them with a small file or burnish them to a shine with the striking surface of a matchbook. If they are not touching when the rack is down, you may be able to bend them slightly to make them contact. Be sure the toaster is unplugged when you are doing this repair and adjustment work.

TOAST WILL NOT STAY DOWN/ TOAST WILL NOT POP UP

1. The bread is lowered in a rack or bread carriage and locked down by a latch that engages the carriage. In normal use the latch may become bent, or it may bind at hinge points

and fail to lock the bread carriage down. On most models the latch-down mechanism is located on the side of the toaster that contains the bread lever and toast controls. Operate the bread lever and watch the action of the latch-down (with the toaster unplugged and the access panel removed). Realign any bent parts and free the latch-down if it's binding.

2. If the toast does not pop up you can suspect that the slide lever which lowered the toast is binding. Operate the slide lever with the toaster unplugged and free up any binding. It is also possible that the latch-down mechanism is broken and not releasing the bread rack, or is binding and holding the bread rack down. Operate the latch-down mechanism and determine if any part is binding or broken.

8

COFFEE MAKERS

You've heard it many times: "There's nothing like a good cup of coffee."

Very true, and sometimes you can even locate a good cup of coffee. Restaurants, as a general rule, serve only a reasonable cup of coffee. Sometimes a better restaurant might serve a better cup of coffee, but then along will come an inexpensive hamburger joint with a really excellent cup of coffee. Sometimes it seems almost more a matter of chance than choice. On any given day conditions are just right (water, heat, bean, brewing time, etc.) and the coffee is excellent. The next day, at the same place or at your own breakfast table, the coffee is only fair or, worse, dishwater.

But you can help control the matter.

HOW THEY WORK

Coffee makers normally brew coffee either by percolation—that is, by circulating water repeatedly through the ground coffee beans—or by dripping water through the ground beans one time.

Percolators may be automatic or nonautomatic. In the latter type, the coffee percolates until you shut it off, and the coffee does not maintain a warm temperature automatically. This type of coffee pot is generally just set on a stove or a campfire and removed when the coffee is ready.

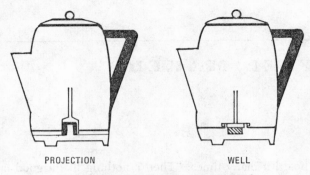

PROJECTION   WELL

TWO TYPES OF COFFEE MAKER HEATERS

An automatic percolator has a heating element that heats water (you should always start with cold water) and perks it up a pipe and into a basket of coffee. The hot water runs through the coffee and back into the pot. Percolation continues until the coffee maker is shut off by a thermostat (which on some units may be preset by a dial to make weaker or stronger coffee). Then a warming circuit takes over and keeps the coffee at drinking temperature until the unit is unplugged. Some have a pilot light that tells when the coffee is ready.

A coffee maker that works by a drip method has two compartments. The hot water is poured in the upper chamber and runs down through the ground coffee into the lower compartment.

A variation of the drip method is the vacuum method, in which the lower compartment is automatically heated and the water is pushed up into the upper compartment by steam pressure inside the lower compartment, then allowed to brew with the ground coffee in the upper compartment. The heat automatically shuts off, and as the lower compartment cools, the coffee is drawn back down through filters, with the coffee grounds being left behind in the upper compartment.

The very latest coffee makers are of the drip type but with a separate reservoir for the water. Cold water is placed in the reservoir, heated, and allowed to drip through a spout into a

COFFEE MAKERS

container of ground coffee. The heated water passes through the coffee, through a filter, and is collected in a serving pot on a warming tray, also built into the unit. From there the coffee is ready to serve.

COFFEE MAKER WON'T HEAT

It is always possible that no current is available at the plug (check the house fuses or circuit breakers) or that the cord or line plug is damaged at either end. This type of plug can be dismantled and repaired if damaged (see Chapter 1).

INDICATOR LIGHT DOESN'T WORK

1. Most modern coffee makers have a light that goes on to indicate when the coffee is ready to drink (on some models the light goes *off* when the coffee is ready). This light can be hooked directly into the warming circuit, coming on when the warming element comes on. If the light doesn't work, the first thing to look for is a burned-out bulb. On most coffee makers, the bulb can be replaced easily by locating the screws holding the base of the unit to the pot, removing them, then changing the bulb in its socket.

2. A defective thermostat can prevent the light from operating. You can check the thermostat in your automatic percola-

tor by removing the pump assembly (the long pipe holding the basket), filling with cold water, then plugging the unit in. Place a thermometer in the pot. For any other type of coffee maker, fill it with cold water and activate the cycle with a thermometer in the pot. The thermostat should open, shutting off the heat source, at a water temperature of about 190°. If it doesn't, you can suspect the thermostat and replace it. If the thermostat is working correctly, check to be sure the warming element is operating. It may not be turning on when the thermostat switch orders it to. You can tell this if the coffee is hot at first, then gradually cools off.

WATER HEATS BUT WON'T PERCOLATE

A defective thermostat can allow the water to heat up but not get quite hot enough to percolate. See the following section for how to adjust the thermostat. If that doesn't work, you will have to replace it.

COFFEE TOO STRONG/COFFEE TOO WEAK

It is quite possible that with the use a coffee maker gets, the thermostat linkage can change and this can change the flavor of the coffee. The coffee maker itself does not know how strong or weak the coffee is, it only knows when the temperature in the pot reaches a certain point. Then it shuts down. If the thermostat changes, it may be waiting too long, or not long enough, before acting. Check the thermostat according to the directions in the section "Indicator Light Doesn't Work."

Some pots have an adjustable thermostat marked for stronger or weaker coffee, which should take care of minor flavor problems. With a fixed, nonadjustable thermostat, you might be able to bend the case slightly to change the setting. You will be able to see the contact point arm or arms. Use two pairs of pliers and bend the case so that the points have more tension on them to increase the heat range. To decrease the heat range, do the opposite. This is a trial and error procedure. If it doesn't work, replace the defective (which it might well be by then) thermostat. When you do, don't tighten it too tight since this can only accomplish the same thing you were trying to do with the pliers, only now you don't want to change the setting on the brand-new thermostat.

COFFEE HAS POOR FLAVOR

If there is one thing that kills more cups of modern coffee-maker coffee than anything else, it is a lack of cleanliness. Coffee pots must be kept scrupulously clean to give the beans a chance to burst into flavorful, untainted coffee. The slightest bit of residue or other soil can ruin a pot of potentially good coffee.

Do not scour the interior of any coffee maker. Most manufacturers recommend that their products be cleaned regularly with hot water and a detergent and rinsed in clear water after *each use*. This may not be realistic, but pots (except for aluminum) should be cleaned periodically with a boiling solution of soda and water; just cycle the solution through the coffee maker if possible, starting with a cold solution and allowing the coffee maker to heat it. You'll be amazed at the difference in flavor if these maintenance procedures are followed.

After the pot has been cleaned and dried, allow it to air.

Some coffee makers can be immersed and some cannot. As a rule of thumb, if the unit which contains the heating elements can be left behind, the rest of the pot can generally be immersed for cleaning. Some coffee makers, however, can be completely immersed since they have a sealed heating chamber. Others cannot. Read the directions with your coffee maker.

Remember, cleanliness will contribute more to a good cup of coffee than almost any other factor.

9

OTHER SMALL APPLIANCES

◆ ELECTRIC IRONS

Sure, many people still use irons, even in this day of "permanent press" fabrics. And, yes, they still have occasional problems with these almost foolproof appliances, especially since most modern irons are also designed to produce steam as a part of their operation.

NOT ENOUGH STEAM

Minerals in the water in the iron can clog the steam ports and foul the steam chamber inside the iron. (Some iron instructions stress that only distilled water should be used.) These deposits are very hard and can restrict or stop the flow of steam. Unclog the ports with one end of a paper clip. For the steam chamber inside, mix a solution of vinegar and water (half and half) and fill the iron. Then turn the setting to "steam" and allow all of the solution to steam out. This should clean the chamber and finish the cleaning of the steam ports.

WORN OR BROKEN HEATING ELEMENT

Earlier electric irons were built with a heating element of nichrome wrapped around a mica support and these could be replaced if they failed. But most modern irons have the heater built into the base casting. Although you can check them for continuity where the wires are attached to the base and possibly locate an open circuit, the heater in the base plate is nonrepairable.

INCORRECT TEMPERATURE

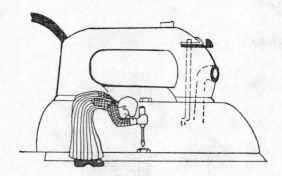

If the iron changes temperature as you change the setting of the thermostat, but the temperature is not correct according to the dial, you may have an iron with an adjustable thermostat. In some of these irons, the thermostat adjustment screws can be reached through a hole in the cover assembly after only partial disassembly. As a general rule, turning the adjustment screws clockwise will decrease the temperature and turning them counterclockwise will increase the temperature. Experiment.

OTHER SMALL APPLIANCES 85

◆BROILERS AND ROTISSERIES

These small portable ovens are popular appliances in the modern kitchen. Most of them consist of a box with one or more heating elements (bottom, top, or both in most cases), a thermostat to control the heat, and sometimes a spit to rotate the food being cooked.

These small appliances generally vary in wattage from a low of about 700 watts to rotisserie ovens large enough for baking cakes with wattages up to about 1,500. All should be plugged into an appliance plug, and if an extension cord is necessary, it should be an appliance extension and not an ordinary lamp extension cord. They should all be used in an area where no combustible material might come in contact with them and where no children can touch them. They become quite hot in operation, so choose one with heat-resistant handles.

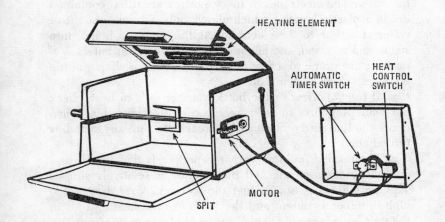

1. These are simple appliances, some of them without even an off-on switch. The heating element heats when the unit is plugged into the wall. If it doesn't heat, almost certainly there is no current in the wall plug, or a connection has loosened behind the cover panel where the wires enter the unit. Unplug it and check for broken or loose wires.

2. If the heating element is broken, it is possible to repair it. Refer to Chapter 1.

◆ VAPORIZERS AND BOTTLE WARMERS

Although they do two different jobs, both of these small appliances have electrodes which project into a water reservoir and as the current passes from one to the other, meeting the resistance of the water, heat is formed. The heat warms a bottle in one appliance and turns water to medicated steam in the other. The electrodes in the vaporizer are often contained inside a plastic housing which allows only a small amount of water at a time to the electrode. As this water is boiled into steam and released, another small amount of water enters. Vaporizers are prone to a few more problems than simple bottle warmers, which seldom fail.

1. If the vaporizer or the bottle warmer fails to heat, check the wall plug first to be sure electricity is available. Then check the line cord and plugs for breaks. Be sure any switch is turned on.

2. The normal problem with a vaporizer is that it will produce little or no steam, and the solution is relatively simple. The most likely reason is that the water in your area is too high in mineral content and that with use the heating rods in the unit have become coated and need cleaning. First, try cycling the unit with one cup of vinegar added to the water. If that doesn't correct the condition, remove the top of the va-

OTHER SMALL APPLIANCES 87

porizer and scrape the accumulated deposits from the rods with a sharp knife. Wash, scrape, and thoroughly clean the inside of any plastic tube, and be certain any water-access hole in the tube is clean and no deposit is blocking a free entry of water. Be sure the electrodes are parallel and not touching each other.

3. If your vaporizer boils too vigorously, emits hot water instead of steam, steams too vigorously, or blows fuses, any several causes can exist. Possibly you have attempted to increase the conductivity of mineral-free water by adding baking soda or borax (nonactivated) and simply added too much. Perhaps the water in the unit is too conductive for other reasons, such as high mineral content or chemically treated or softened water. Empty the water and half fill with distilled water, rain water, or water from your freezer or refrigerator defrost pan, then add faucet water to the "full" mark.

Most vaporizer manufacturers issue a repeated warning on instruction sheets that you should never aim the steam jet at a patient. It is possible, under certain overheating conditions, for the steam jet to emit a stream of superhot water.

♦ ELECTRIC FRYING PANS
AND COOKERS

The simplest way to use electricity in an appliance is to convert it to heat. If electricity flows through any wire or ribbon made of metal that is somewhat resistant to the flow, heat is generated. This conversion from electricity to heat requires no moving parts at all, though we often add control parts in the form of switches, thermostats, and other convenience additions.

Generally the wire that heats up is a combination of nickel and chromium, or nichrome. Nichrome can operate at temperatures over 2,000° without melting or other heat damage.

Most modern electric frying pans and other cooking pans combine a plug-in switch and thermostat control with a heating element built into the skillet. The skillet can be immersed in water for cleaning after the switch section is unplugged.

The heating element–pan section of these appliances is

OTHER SMALL APPLIANCES

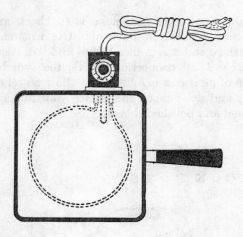

nonrepairable. If something goes wrong, the whole pan is generally replaced.

The thermostat-switch section can be repaired, but if the temperature range has shifted far from where it should be it might be wise to replace this as well. Most of these plug-in units contain simple bimetal thermostats, using the principle that different metals expand at different rates when heated and contract at different rates when cooled. Two thin strips of metal, each with a different but known expansion and contraction rate, are fused together. If one end of this double strip is fastened in place and heat is applied to the center, one metal expands more than the other, pushing or warping against the other, and the strip will bend.

This bending action can be used to open and close a switch as heat is applied. When enough heat has been applied, the strip bends and opens a switch. As the strip cools down, it bends back and makes contact again, which closes the switch. If the switch controls current to a heating unit, and if the heating unit is the source of heat for the bimetal switch, a thermostat has been created.

The main repair you can make is to check and burnish clean the contact points in the switch. Use a matchbook-cover striking surface and polish them clean and flat. Also clean the terminals, check all connections inside the switch unit, and add a drop of oil if rust is found (but allow no oil on the contact points and don't touch them with your fingers after you have cleaned and polished them).

♦ CROCK POTS

One of the most modern appliances of all has been around—nonautomatic except for grandmother's skill with the cookstove—for hundreds of years. Sometimes they are called slow cookers, but most people refer to them simply as crock pots. This is because they are made of crockery (and for that reason must be handled with care, for they will break if dropped). The difference between grandmother's method of slow cooking and the modern way is that the newfangled crock pots have low-wattage heat *all around* the cooking food and not just beneath it. In fact, most modern crock pots do not heat from the bottom at all but only from around the sides. This eliminates scorching of food and there is less chance of overcooking and shrinking of meats.

OTHER SMALL APPLIANCES

Most modern crock pots have two heat settings, low and high, and a few now have an automatic switch that turns the heat from high to low after a certain time period. In both types, heat is controlled by a thermostatic switch which allows current to flow through heating coils in the crockery. Few slow cookers may be submerged in water for cleaning, so read the directions with your unit. Abrasive cleaning compounds should never be used on crockery, and stains should be carefully removed only with a crockery stain remover, never with other commercial cleaners, scrapers, or metal spatulas.

Here are three tips on crock pots:

1. Do not put in frozen or very cold foods if the pot has been preheated or is hot to the touch.

2. Do not use your crock pot as a storage container in the refrigerator.

3. Do not wash your crock pot after cooking until it has had a chance to cool off.

The crock pot is an appliance that should last for years without trouble, thanks to its enclosed heating element, its simple thermostat switch, and its low voltages.

Refer to the preceding section on electric frying pans and cookers for further information and the handling of problems.

◆SPACE HEATERS

There are two general types of space heater in use today, both for the heating of smaller areas. These are the bowl type and the convection type.

The bowl-type heater gets its name from the bowl shape of the metallic reflector. It is generally mounted on a base and protected by wire guards to keep hands (and flammable drapes, etc.) away from the heating element, consisting of a resistance wire wound around a cone-shaped insulator. The heat is radiated much the same way as a beam of light radiates from a flashlight.

Convection-type heaters have heated wires that heat the room by both radiation and convection, either by natural draft or by fan-forced air.

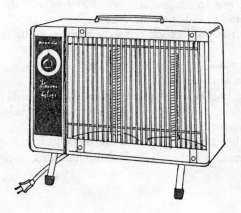

1. If the heating element of any space heater does not work, first check the electric plug in the wall and the line cord and plugs to the heater. If these are all right, suspect an open circuit in the heater. Before repairing or replacing the heating element, be sure the unit is unplugged.

It is possible that an open circuit in a bowl-type heater can be found and repaired (see the section "Heating Elements" in Chapter 1), but generally its heating unit is more easily replaced than repaired. Simply unscrew it like a light bulb and replace it with a new one of the same type. Be sure to screw the unit in solidly. These units carry quite a bit of cur-

OTHER SMALL APPLIANCES

rent, and if one is loosely inserted, an arc can result which will weld the unit in place in its socket, making it almost impossible to remove next time.

In convection-type heaters also, the most common problem is a broken heating-element wire. Be careful not to damage the asbestos insulating washers when removing the element from its frame. Replace anything you damage with a new part.

2. Replacing the fan in a forced-air convection heater is quite simple. Remove the old one and put in the new one the same way. Remember that fan blades are adjusted at the factory. Before you turn the fan on, be sure the blades are all traveling in the proper path and that nothing has been bent, before operation.

3. If the plug should become too hot (uncomfortable to touch and not just a normal warmness) in any electric heater, suspect a poor connection at the wall outlet. Space heaters require a large amount of electric power and they are often operated all night long. Not only will a poor wall connection overheat the plug, it also represents a potential fire hazard. Remove the fuse or trip the circuit breaker, remove the cover plate from the wall outlet, and check and tighten the house wire connections. Carefully examine the prongs in the appliance plug. Be sure they are firmly attached to the plug

body. If any sign of burning or other deterioration is noted, replace the plug.

4. If the heating element seems to be working but it doesn't put out a sufficient amount of heat, be sure the heater is not being required to heat too large an area. Also be sure the chromed reflective surface is clean and shiny. If not, clean and polish it with a soft cloth and metal polish. It is surprising how a clean reflector will increase the heat from a space heater.

◆ HAIR DRYERS

The operation of most hair dryers is very similar to the operation of space heaters. That is, air is heated by a heating unit and moved by a fan to where it is needed. Most hair dryers are constructed so that the heating element will not turn on if the motor will not turn on (otherwise a heat and fire hazard could result).

1. Check the electric plug in the wall to be sure electricity is reaching the unit. Also, suspect an open circuit breaker or fuse.

2. It is possible that the heating element is broken. Refer to Chapter 1.

3. It is possible that the thermostat switch is defective, but with this small appliance it could be more practical to replace the switch than to take the time to try to repair it, if it can be replaced. Or it might be better simply to replace the entire dryer.

◆ GARBAGE DISPOSERS

Few homes are now without the convenience of a garbage-disposal unit, which will get rid of at least the softer kitchen

OTHER SMALL APPLIANCES

wastes, and some of them will eat up bones with a happy hum. Provided the switch is working and electricity is reaching the unit, there are two other problems that most often develop.

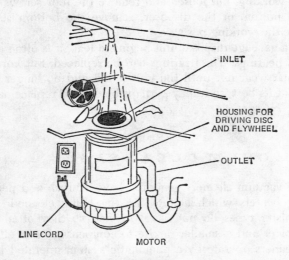

INLET

HOUSING FOR DRIVING DISC AND FLYWHEEL

OUTLET

LINE CORD

MOTOR

(SWITCH IS MOUNTED AWAY FROM UNIT CONVENIENT TO USER)

1. If the motor hums but the unit does not operate, suspect that waste has jammed the cutting and grinding blades. This can happen if the unit was shut off too quickly after the last use, or if something has been put into it that is too hard, or if too much has been put in at once. Turn off the switch, then use a broom handle or other wooden rod in the inlet of the unit to free the blades. Be sure you run cold water through the unit when it is operated.

2. If the unit does not operate and the motor does not hum, suspect the circuit breaker (usually located on the bottom of the disposer). Press the red button to reset the motor, then try the main switch again.

SIMPLE APPLIANCE REPAIR

3. If you cannot free the blades from above, and if the unit continues to jam when switched on and trips the breaker, it is possible to remove the bottom section of the disposer without removing the entire unit from the sink. You'll find tight quarters for working, but loosen and remove the four screws holding the bottom of the disposer, remove the bottom section, and free the working parts.

4. If a garbage-disposal unit begins to leak, it is often a sign of worn bearings. The bearings can be replaced, but normally they don't wear out until the unit is old and no longer functioning at its best; so it may perhaps be due for replacement.

♦ VACUUM CLEANERS

Upright vacuum cleaners generally have a brush and perhaps rotating beaters which agitate the rug being cleaned. Tank and canister types do not have brushes, so they often have more power and a smaller nozzle to compensate. But all vacuum cleaners operate by a pneumatic system operated by an electric motor. The system sucks in air from the outside, passes it through a filter bag to collect dust and other debris, then blows it out again.

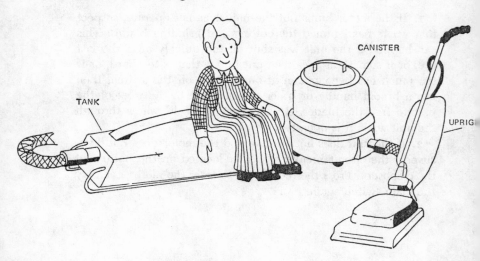

OTHER SMALL APPLIANCES

1. By far the most common problem with vacuum cleaners, assuming the unit itself is functioning, is too much debris in the collecting tank. Vacuum cleaner bags should be emptied or replaced when only *half* full, or even less. Many people suppose that the unit will continue to operate efficiently until the bag is *full*.

2. The unit must have a clear air path from outside through the collecting bag to outside. If the cleaner is not lifting dust and debris as it should, check the air inlet and any pickup tubes to be sure no debris has been caught and lodged inside, blocking a clear flow of air.

3. If an upright cleaner's efficiency drops, and the bag is not full, check the belt which drives the brushes and the beater (on the same drum as a general rule). The belt can be easily replaced by following the path of the previous one (usually marked plainly on the castings).

4. It is not difficult to replace the motor in any of these units (replace with exactly the same type) if it should eventually fail.

◆ ELECTRIC CAN OPENERS

Isn't science wonderful? We don't even have to twist to open a tin can anymore. Just pop the can into one of these units, push a handle, and the thing spins itself open with a clean cut. You'll have to be careful when you remove the open can from the opener or you'll spill juice, but it still beats an old-fashioned hand-operated model.

1. Be sure the unit is plugged into a live circuit. Then if it fails to operate, check the off-on switch that is generally a part of the control handle. Check the mechanical linkage to be sure that depressing the handle is activating the switch. Replace any defective parts.

2. If the can opener is too noisy, it is possible the gear train needs lubrication. Remove the case, inspect and lubricate if needed, but use a good gear grease that won't heat and run.

3. Sharpening the cutting blade often takes more time and results in less efficiency than merely replacing it. A couple of dollars and fifteen or twenty minutes is all it will take.

4. One of the major problems with a can opener, a problem that will appear often throughout the life of the unit, is that it will drop cans and tops because the cutting area is dirty and crusted with old food debris. Clean this area regularly with detergent and a toothbrush (and, if necessary, use a knife—carefully—to cut and scrape hardened encrustations). You'll be surprised how this will renew the action.

◆ **BLENDERS AND MIXERS**

Blenders and mixers are similar in operation. In both appliances a motor drives a tool that processes food in a cer-

OTHER SMALL APPLIANCES

tain way. In a blender the motor and tool are below the food container; in a mixer they are above. In a blender the food container is generally held in place by four ears; the cutting blades are in the container and are coupled with the motor shaft when the container is placed in position.

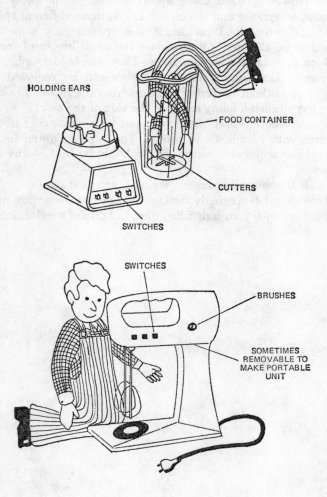

1. With both appliances be sure the plug is operating and in a live circuit. The motors in these appliances, because they are rated to do heavy work and are not used as frequently or for as long a period as other appliance motors, rarely give any problem.

2. However, since these appliances use a universal type motor, a versatile unit which can run at many different speeds under different load conditions, the brushes may sometimes need to be changed. (Brushes are the small blocks of carbon which contact the turning rotor. They are held in place by springs.) On many units the brushes can be removed and replaced without removing the case from the motor. Look for the large brush-holding screw in the side of the case.

3. Bent cutters and beaters account for a continuing trouble source with blenders and mixers. You can straighten blades, but if the shafts are bent it will probably be best to buy new ones.

4. If the motor becomes noisy and erratic, it may only need lubrication. Oil sparingly through oil holes or according to instructions with your individual unit, but do not overlubricate.

INDEX

Adjustable wrench, 1
Agitator (washing machine), 37, 47
  won't operate, 43
Appliance plugs (male and female), 8–10
Appliance repair, 1–12
  checklist, viii–x
  general procedures for, 1–12
  suggestions for, vii-viii
  tools to use, 1

Belts (washing machine), 42
Blenders and mixers, 98–100
Bottle warmers, 86–88
Bowl-type space heaters, 91–94
Broilers, 85–86
Bypassing (water heater), 27–28, 31

Can openers, electric, 97–98

Channellock pliers, 1
Circuit breakers, 2–5
  current overload, 3
  increasing circuit load rating, 5
Claw hammer, 1
Clogged drain (washing machine), 45
Clogged drain hose (dishwasher), 20
Clogged filter screen (dishwasher), 18
Clogged filter screen (washing machine), 39–40
Clogged inlet valve (washing machine), 41
Clogged lint filter (washing machine), 43–44
Clogged strainer screen (dishwasher), 19–20
Clothes dryers, 49–56
  checking, 50
  clothes don't dry, 53–54
  continues to run with door open, 55
  how they work, 49
  noisy operation, 55–56
  runs with door open, 55
  too much heat, 54

won't heat, 51–53
won't run, 50–51
Coffee makers, 77–82
   cleanliness, 82
   coffee has poor flavor, 81–82
   coffee too strong / too weak, 81
   how they work, 77–79
   indicator light doesn't work, 79–80
   water heats but won't percolate, 80
   won't heat, 79
Common screwdriver, 1
Compressor noise (refrigerator or freezer), 65
Contact points (toaster), 74
Convection-type space heaters, 91–94
Cookers, electric, 88–90
Crimped inlet hose (washing machine), 39
Crock pots, 90–91
   tips on, 91

Detergent, dishwasher, 13, 15, 59
   manufacturer's instructions and, 20–21

Diagonal cutting pliers, 1
Dishwashers, 13–23
   checking, 15
   how they work, 13–15
   leaks, 16–17
   won't clean, 20–22
   won't dry, 22–23
   won't empty, 19–20
   won't fill, 17–19
Dollar-bill test, 59–60, 64, 67
Drip-type coffee makers, 78–79

Electric can openers, 97–98
Electric cookers, 88–90
Electric frying pans, 88–90
Electric irons, 83–84
   incorrect temperature, 84
   not enough steam, 83–84
   worn or broken heating element, 84
Electric shock, ix–x
Electric water heater
   heating elements in, 31
   "singing" noise from, 35
   See also Water heaters
Evaporator coil (refrigerator or freezer), 59, 64

INDEX

Fan drive belt (clothes dryer), 54
Female appliance plug, 8–10
Flashlight, for examining toaster, 72
Freezer. *See* Refrigerators and freezers
Frying pans, electric, 88–90
Fuses, 2–5
  current overload, 2–3
  increasing circuit load rating, 5
  locating the short circuit, 5
  slow-blow type of, 4

Garbage disposers, 94–96
Gasket leak (refrigerator), 64
Gas leak (water heater), 30

Hair dryers, 94
Heating elements, 11–12
Hot water tanks. *See* Water heaters

Indicator light (coffee maker), 79–80
Inlet valve (washing machine), 40, 41
Irons, electric, 83–84

Jammed door switches (dishwasher), 18–19

Kinked inlet hose (washing machine), 39
Knife, 1

Leaks
  dishwasher, 16–17
  refrigerator or freezer, 64
  washing machines, 44–45
  water heater, 32–33
Levelers (washing machine), 47–48
Line plug (coffee maker), 79
Lint filter (washing machine), 43–44

Male appliance plug, 8–10
Mixers. *See* Blenders and mixers

Needle nose pliers, 1
Noise
  clothes dryer, 55–56
  refrigerator or freezer, 64–66
  washing machine motor hum, 42–43
  from the water heater, 34–35

Odors (refrigerator or freezer), 68

Percolators. *See* Coffee makers
Phillips screwdriver, 1
Pilot light
  coffee maker, 78
  water heater, 29
Pipe system (water heater), 32
Pipe wrench, 1
Pliers, 1
Plugs, replacing, 8–10

Refrigerant liquid, 58
Refrigerators and freezers, 57–68
  checking the controls, 59
  excessive power consumption, 66–68
  how they work, 57–58
  noisy operation, 64–66
  not cold enough, 63–64
  odors, 68
  too cold, 61–63
  too much frost, 59–60
  water leaks on the floor, 64
  won't run, 60–61
Rotisseries, 85–86

Sacrificial anode, 26, 28, 34
Screwdrivers, 1
Sediment (in water heater tank), 34
Service policy, viii-ix
Short circuit. *See* Circuit breakers; Fuses
Slow-blow fuse, 4
Soldering iron or gun, 1
Soldering wires together, 8
Solenoid, 40
Space heaters, 91–94
Splicing and connecting wires, 5–8
Spray arms not rotating (dishwasher), 22

INDEX

Stronger or weaker coffee,
   adjustable thermostat
   for, 81

Terminals, 5–8
Thermostat switch
   (refrigerators and
   freezers), 61
Toasters, 71–75
   cleaning, 72–73
   how they work, 71
   toast does not brown / is
      too brown, 73–74
   toast will not stay
      down / will not pop
      up, 74–75
Tools, list of, 1

Vacuum cleaners,
   96–97
   emptying bags, 97
Vacuuming the clothes
   dryer, 52
Vaporizers, 86–88

Warranties, viii
Washing machines, 37–48
   agitator won't operate,
      43
   checking, 38–39
   excessive vibration,
      46–48
   how they work, 37–38
   leaks, 44–45
   lint filter not working,
      43–44
   too much water, 41–42
   won't empty, 45–46
   won't fill, 39–40
   won't run (but motor
      runs or hums), 42–43
Water heaters, 25–35
   checking, 27–28
   how they work, 25–27
   leaking, 32–33
   noises, 34–35
   not enough hot water,
      31–32
   won't heat, 28–31
Wire connectors, 5–8
Wrench, adjustable, 1